Food Chain Kings

CROCODILES, ALLIGATORS AND THEIR FOOD CHAINS

CHERITON
CHILDREN'S BOOKS

Published in 2023 by **Cheriton Children's Books**
1 Bank Drive West, Shrewsbury, Shropshire, SY3 9DJ, UK

First Edition

Author: Katherine Eason
Designer: Paul Myerscough
Editor: Elise Harding
Proofreader: Hayley Bennett
Illustrator: Martyn Cain

Picture credits: Cover: Shutterstock/GoodFocused; Inside: p1: Shutterstock/Worldswildlifewonders; p4: Shutterstock/Dave Montreuil; p5: Shutterstock/Dennis Molenaar; p6: Shutterstock/Worldswildlifewonders; p7t: Dreamstime/Sohadiszno; p7b: Shutterstock/Roger de la Harpe; pp8-9: Shutterstock/Sergey Uryadnikov; p10: Shutterstock/Vladimir Turkenich; p11: Dreamstime/John Anderson; p12: Shutterstock/Johan Larson; p13: Shutterstock/Goodluz; p14: Shutterstock/Sweetangel74; p15: Shutterstock/Svetlana Foote; pp16-17: Shutterstock/Thierry Eidenweil; p18: Shutterstock/Nima Nata Lia; p19: Shutterstock/Andrew Nicholas White; p20t: Shutterstock/Stephen; p20b: AdobeStock/Paul Wolf; p21: Shutterstock/Four Oaks; p22: Shutterstock/Roger de la Harpe; p23: Shutterstock/Wildestanimal; p24: Shutterstock/FloridaStock; p25: Shutterstock/Nick Greaves; p26: Shutterstock/WeStudio; p27: Shutterstock/Catchlight Lens; pp28-29: Shutterstock/Nikolay 007; p30: Shutterstock/Catmando; p31: Shutterstock/Vedderman123; p32: Shutterstock/YuRi Photolife; p33: Wikimedia Commons/Ghedoghedo; p34: Wikimedia Commons/Daderot; p35: Shutterstock/Daniel Eskridge; pp36-37: Shutterstock/Michael Rosskothen; p38: Shutterstock/Iofoto; p39: Shutterstock/De Visu; p40: Shutterstock/Photomika-com; p41: Shutterstock/Mikhail Blajenov; pp42-43: Shutterstock/Vaclav Matous.

Printed in China

CONTENTS

CHAPTER 1

KILLER KINGS

Crocodiles and alligators are meat eaters. They hunt other animals that live in their ocean, river, or **wetland** homes. Their **prey** includes fish and animals that live in the water or come to the waterside to drink.

King Crocs and Awesome Alligators

Crocodiles are mostly found lurking in the waters or along the banks of **swamps**, marshlands, lakes, and rivers. They thrive in warm, **tropical** regions of Africa, Asia, Australia, and a few are found in North and South America. Crocodiles are the top **predators** where they live. Some are large and can be very dangerous. Alligators are at home in the swamps and wetlands of the southeastern United States. They are most common in Louisiana and Florida, where you may often see one sunning itself on the edge of a river or lake.

Fish are a favorite food for many crocodiles.

Chain Champs

Crocodiles and alligators are part of ocean, river, and swamp food chains. A food chain is a series of animals and plants that are connected because they depend on each other for food. Each animal in a food chain is linked to the animals and plants below it. Crocodiles and alligators are at the top of many food chains because they hunt fish and other animals, which in turn feed on smaller animals or plants. Plants are at the bottom of most food chains.

Nutrias are a favorite food for alligators in Louisiana.

On the Menu?

Although their ferocious-looking teeth make crocodiles and alligators appear scary, attacks on humans are rare. But they do happen … Crocodiles are more likely to attack humans than alligators. Even then, they mostly only attack people if they're hungry, they're defending their **territory**, they're defending their young, or they intend to attack a pet that was near a person.

Links in the Chain

Food chains are made up of producers, consumers, and decomposers. Animals that eat other animals are called secondary consumers, and the plant-eating animals they eat are called primary consumers. The plants that primary consumers eat are called producers. Decomposers break down the waste produced by consumers and producers.

Watery Hunters

Crocodiles and alligators are reptiles. These are animals that breathe air and have a scaly skin. Reptiles warm themselves by taking in heat from their **environment**. Alligators and crocodiles belong to a reptile animal family called crocodilians. Crocodilians live in hot, wet places and include caimans and gharials as well as crocodiles and alligators.

Two Types of Killers

There are two types of alligators —American alligators and Chinese alligators. Only a few Chinese alligators still survive in the wild. American alligators share their wetland home with caimans and American crocodiles.

Crocodile Club

There are 14 species of crocodile. Some live in the warmest parts of the Americas and the Caribbean. The American crocodile is the most widespread. There are a small number in southern Florida, but most are found in southern Mexico, Central America, the Caribbean, and northern South America.

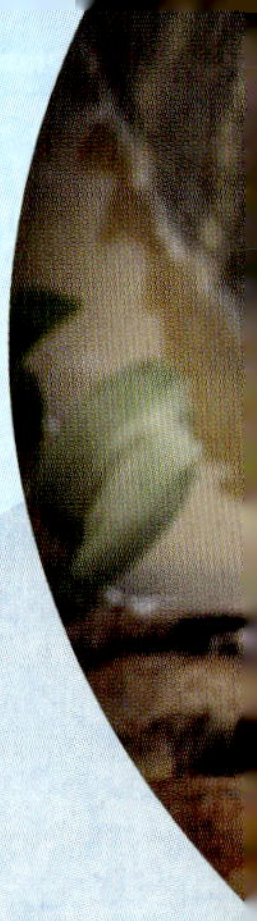

A crocodile has a long snout!

Long or Short

You can tell a crocodile from an alligator by its **snout**. A crocodile has a long narrow snout, while the snout of an alligator is short and broad.

Most caimans live in tropical rivers and swamps in Central and South America.

Biggest Killer Croc

Of the crocodiles that live in Africa, the Nile crocodile is the most famous. It lives in a wide variety of watery **habitats**, from sub-Saharan countries to the tip of South Africa. The remaining species of crocodiles are found in southern Asia and Australian waters. Nile crocodiles are the largest crocodiles in Africa. They lurk in rivers and lakes, and snatch animals that come to the water's edge to drink. Their prey includes large animals such as wildebeests, along with smaller animals such as antelopes.

Few animals are safe from a Nile crocodile's teeth!

FEROCIOUS HUNTER

The Nile crocodile is one of Africa's most feared and aggressive hunters. This giant of a crocodile grows up to 16 feet (4.8 m) long.

This Nile crocodile has caught a wildebeest with its mighty bite.

Nile crocodiles can bite down on prey with a force eight times more powerful than that of a great white shark! Their supremely powerful jaws and sharp teeth mean they can take down almost any prey. Nile crocodiles mainly eat fish that they catch as they float in the water. But they can also snap those jaws fast and hard on animals of almost any size that stray too close to the water's edge.

More Than One Croc

Where in the World?

Until recently, scientists thought that the **range** of the Nile crocodile was much bigger and covered much of Africa. They now know that this is not the case. The very similar-looking but smaller crocodiles found today in western Africa are actually a different **species**, known as the West African crocodile or sacred crocodile.

In the past, the ranges of the two crocs overlapped, but scientists think that the more aggressive Nile crocodile forced the sacred crocodile out of the Nile. It could then exist only in less-suitable habitats, ranging from woodland and hills to coastal lagoons in western Africa. As a result of **climate change** and hunting, the sacred crocodile is now **endangered**.

IN THE FOOD CHAIN

Nile crocodiles eat mainly fish but will also devour larger creatures such as zebras, small hippos, and wildebeest. Young Nile crocodiles feed on small animals lower down the food chain, such as insects and small fish.

Nile crocodile food chain

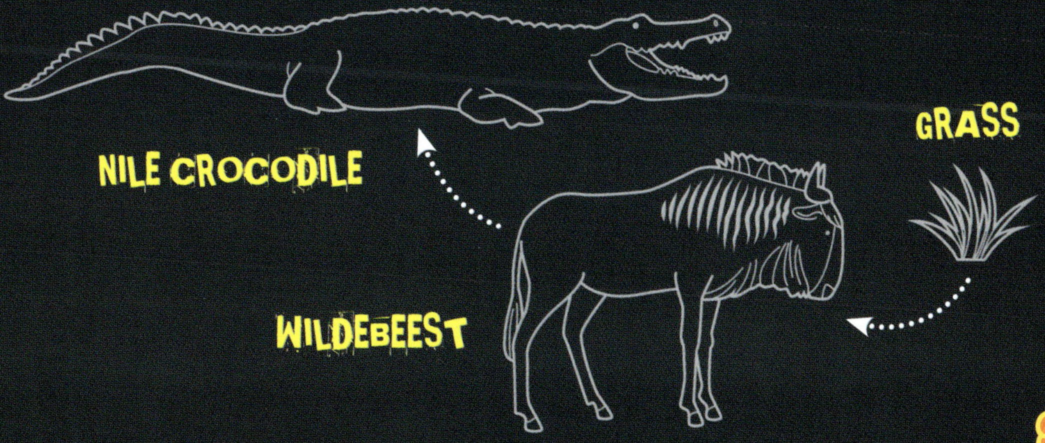

CHAPTER 2

DESIGNED TO KILL

Adult crocodiles and alligators are top predators because they are perfectly designed to hunt and almost impossible to attack. They are fast swimmers with incredibly strong jaws that they use to grab their prey. Once they have caught their prey, the hunters' teeth tear it apart.

Super Senses

In addition to their weapons, crocodiles and alligators have a good sense of smell and great eyesight and hearing. They use all of these senses to find prey. They also have another extraordinary sense. They can "feel" **vibrations** and slight movements in the water, which tell them when another animal is close by. Crocodiles are even more sensitive to vibrations than alligators—their jaws have a stronger sense of touch than the fingertips of a human! In still water, a crocodile can feel the ripples from just a single drop of moving water.

Crocodiles use their mouths to feel for and snap up even small prey.

Supersized and Deadly

American alligators grow up to 15 feet (4.6 m) long and weigh up to 1,000 pounds (454 kg). An average-sized adult alligator is more than twice as long and more than four times as heavy as an average human. Crocodiles vary greatly in size. The dwarf crocodile grows to only about 6 feet (1.7 m) long. At the other end of the scale are the biggest crocodiles in the world. They are the Nile crocodile of Africa and the saltwater, or **estuarine**, crocodile from Australia. These mighty crocodiles both reach lengths of up to 20 feet (6 m) and weigh about 2,200 pounds (more than 1,000 kg). They are the biggest reptiles alive on the planet today, and what's more, they can live until they are 80 years old!

Armor-Plated Killers

Crocodiles and alligators are protected by their thick, scaly skin, which is like a suit of armor. No other animal can pierce the skin, which makes the predators almost impossible to overcome.

Alligators have a big head, four short legs, and a long, strong tail.

Sneaky Hunters

One reason why alligators and crocodiles are such successful hunters is that they are not in a hurry! They can lie in wait for hours with most of their body hidden beneath the water. They do not go looking for prey. Instead, they wait for their dinner to come by.

Sinking Low

An alligator can sink below the surface for up to an hour! An extra, see-through eyelid protects each eye and allows the alligator to see under water. One flap closes its nostrils. Another flap, called the glottis, closes off its throat. This allows the alligator to grab prey without swallowing water.

Hidden Beneath

Only the top of the alligator or crocodile's head shows above the surface of the water, so its eyes can watch for prey and its nostrils can breathe in air. The hunters wait for prey to come near, either in water, on the land alongside the lake or river, or even in the air above their head. Then, when prey arrives, the predators go into action!

A crocodile's eyes are on top of its head, so that it can see all around without moving its head. It waits for unsuspecting prey to swim by.

An alligator hides among plants at the side of the river. It waits for prey to come within reach in the branches or in the water.

Smart Hunter

Crocodiles are very smart in the ways that they hunt. They watch out for land animals that come to drink or bathe at a particular spot. Then, they move to that spot, sink under the water, and wait. Crocodiles are very patient—they can stay still and quiet for hours while they wait at the edge of water where thirsty animals gather. Then, with a sudden lunge forward and up out of the water, a crocodile grabs its prey in its jaws. Once a crocodile has caught its prey, it stabs its sharp teeth into the animal so there is no escape.

In for the Kill

Alligators and crocodiles attack suddenly. If they sense prey in the water, they may just open their huge mouths and swallow it. They may also sink below the surface to grab larger prey from below. If an animal is on the riverbank, the alligator or crocodile moves closer, then throws itself out of the water to snatch its prey.

Most adult buffalos are safe from crocodiles, but baby or sick buffalos are not.

Drag and Drown

When the hunters grab a bigger animal, they drag it into the water. There, they roll over with their prey and hold it under the surface until the animal drowns. Then, the predators use their powerful jaws to bite into the animal's body and rip off a chunk of flesh to swallow.

Not Fussy Eaters

Crocodiles will eat almost any animal that they meet. No meal is too small if it doesn't take too much energy to catch. At times when food is scarce, some crocodiles will even eat snails. However, crocodiles mainly catch and eat fish, turtles, birds, and other small animals. A crocodile is not a picky eater and will also steal prey that other animals have killed.

The Death Roll

When a crocodile catches a large prey animal, it holds the animal tightly underwater in its strong jaws and fierce teeth. Then, it quickly spins and rolls over, again and again. This death roll makes the large prey animal dizzy, which makes it more difficult for it to escape. It also causes chunks of the prey to fall off, which the crocodile then swallows whole.

Buffalo Battle

The Nile crocodile is large enough to take on huge water buffalos. A Nile crocodile usually hunts old, tired buffalos or young buffalos rather than healthy adults. In a battle with a large, fit adult buffalo, the crocodile can come off worse! A buffalo has large, heavy, and curved horns. If the buffalo can push them beneath a crocodile, it can stab the predator in the belly, which is protected only by smooth, leathery scales that are not as tough as those on the back and head.

Snake Fight

Alligators catch fish, turtles, water snakes, and small mammals in the water. Some water snakes, such as the Burmese python, are large enough to fight back. Burmese pythons are moving into alligator wetland territory, but so far alligators have been able to stop the invading snakes from taking over their territory.

Even young alligators are not safe from adult alligators! This young alligator is about to be eaten by an older alligator.

DEADLY HUNTER

The American alligator has huge and amazingly powerful jaws.

An adult American alligator is grayish-black on its back with a light-colored belly. It can measure up to 15 feet (5 m) long. Its massive and very strong tail accounts for about half of this length! It can weigh as much as 1,000 pounds (0.5 mt). Females are usually smaller than males.

The American alligator is a dangerous and deadly hunter. This top predator fears no other animal, except, perhaps, other American alligators. When hunting, an American alligator rests with its eyes and nostrils just above the water. When prey is close enough, the alligator snaps open its powerful jaws and clamps down on its victim with its 80 strong teeth. American alligators capture just about any animal they find in the water or at the water's edge.

Where in the World?

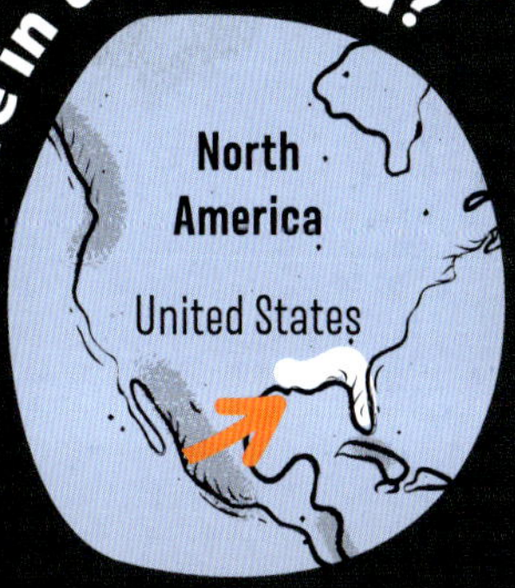

A Threatened Home

American alligators live only in the southeastern United States. Most of them are found in Florida and Louisiana, where they live in rivers, lakes, ponds, swamps, bayous, and marshes. You're most likely to see them in slow-moving rivers. They can only cope with being in salt water for short periods of time. Around 200,000 American alligators live happily in the Everglades National Park, a wetland reserve that stretches across central and southern Florida.

The area in which American alligators live today is getting smaller. The wetlands where they live and hunt are being drained to create land for homes and golf courses. The water they rely on is also being damaged by chemicals and other forms of **pollution** that wash off the land through activities such as farming. When pollution enters the alligator's watery habitat, it affects all the living things in it, from plants to animals. That causes untold damage to food chains.

IN THE FOOD CHAIN

American alligators eat fish, birds, snakes and **mammals** that come to the water's edge. Younger alligators eat smaller animals in the food chain, such as insects, shrimps, snails, small fish, tadpoles, and frogs.

American alligator food chain

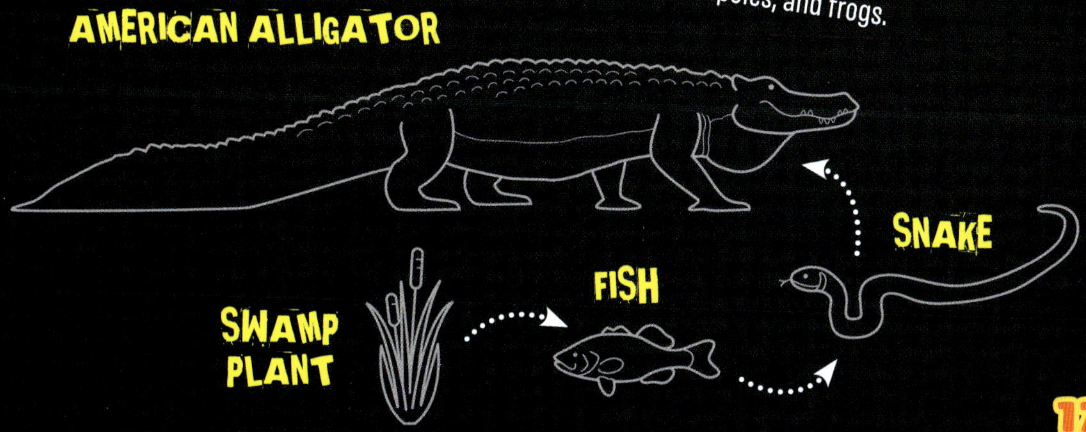

CHAPTER 3

KILLER BODIES

Crocodiles and alligators use speed to catch their prey. They can move quickly in the water and on land. They swim by moving their strong tails. The big, flat tail has a large surface area that pushes against the water as the reptile moves from side to side. This movement makes the crocodiles and alligators swim along in an S-shaped pattern. They also use their feet, which are partly-**webbed**, to push them through the water.

Speedy Hunters

A crocodile can swim at speeds of up to 18 miles per hour (29 km/h) over short distances. An alligator swims at speeds of up to about 20 miles per hour (32 km/h). When compared with humans, who can swim on average at about 1.8 miles per hour (3 km/h), it's not difficult to understand why you should avoid taking a dip in the predators' waters!

A running crocodile is a terrifying sight.

When moving quickly into the water from a bank, crocodiles slide on their bellies and push themselves forward with their feet.

Chasing Down Prey

Alligators and crocodiles run quickly on land by pushing against the ground with their legs and feet and twisting their bodies from side to side. Alligators can run at speeds of up to 9 miles per hour (14 km/h), which is more than twice as fast as a quickly walking person can move. Crocodiles are only a little slower. You would have to run to get away from either an angry alligator or crocodile!

Land to Water

If a crocodile or alligator is on land and needs to escape into water, or spots a prey animal in it, they have another way of moving. They drop down onto their bellies, lift their front legs, and push off with their back legs to slide speedily from land to water.

NOWHERE IS SAFE

Even birds and animals hiding in trees that hang over the water are not safe from alligators and some crocodiles. Their tails are so strong they can push the hunters upward, so they can jump out of the water to catch a bird flying close to the surface, or to snatch prey from low branches above.

Armed and Dangerous

The jaws and teeth of an alligator or crocodile are its most terrifying weapons. Its jaws are so powerful they can crush bones and the shells of turtles! Although the muscles that shut a crocodile or an alligator's jaws are extremely strong, the muscles that open them are very weak. This is why a person can hold a crocodile or alligator's jaws shut with just one hand.

This man is holding an alligator's jaws closed. The alligator's muscles are not strong enough to open the mouth and bite him.

Teeth Factory

An alligator's long jaws contain up to 80 teeth, which it uses to grip and then crush its prey. A healthy crocodile can have as many as 60–110 teeth, depending on the species. When a crocodile or alligator tooth becomes worn down, it falls out and a new one grows in its place. By the end of its life, an alligator or crocodile may have used 3,000 razor-sharp teeth.

Alligators can crush a turtle shell like an egg!

TURTLE CRUNCHER

Turtles are well-protected from most water animals, but not alligators and crocodiles! An alligator or crocodile swallows small turtles, but also catches and crunches up larger ones. Turtles feed mainly on fish, although some eat only plants.

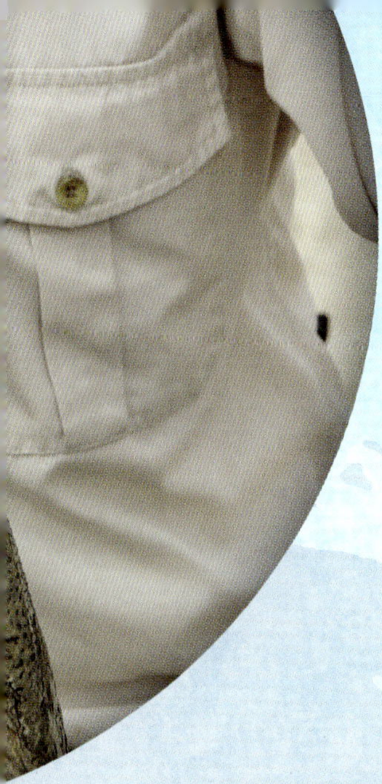

The Strongest Bite

Lions and tigers may be famous for having a hefty bite, but the award for the strongest bite in the world goes to the crocodile. Saltwater crocodiles have the strongest bite of any animal on Earth. They can clamp their large jaws shut with a force of around 4,400 pounds (2,000 kg). That's almost twice as strong as a great white shark! In addition, saltwater crocodiles can snap their jaws closed in just 50 milliseconds. That's six times faster than humans can blink their eyes. No wonder these reptiles are such successful predators.

Crocodiles have pointed, cone-shaped teeth that can grab onto and grip wriggling prey.

Super-Tough Armor

Along with having a tough, scaly skin, an alligator is also protected by bony plates that form a suit of armor under its skin. It is almost impossible for another animal to pierce an alligator's armor or the ridge of bones along its back.

Sensitive Killer

An alligator's skin is useful in other ways, too. Although it is tough, the skin is sensitive to tiny movements in the water. With its bumpy skin and dull brown or gray color, an alligator is well-**camouflaged** as it lies in wait for its prey. It looks just like a log floating below the surface of the water, until it bursts to life and attacks its victim!

War Wear

The scales on a crocodile's back and tail are raised, hard, and bumpy. The hard bumps create the armor plating that protects the predator from injury during fights with other crocodiles or alligators. It also protects them from the heavy hooves or horns of large prey animals that they attack. However, prey animal weapons have little chance of piercing through the armor plating on a crocodile's back.

A crocodile is covered with tough scales that fit closely together.

A crocodile's scaly skin even helps it swim without creating ripples that might give it away.

Scales for Swimming

The raised and bumpy scales along the tail of a crocodile or alligator help the animal swim more **efficiently**, by channeling water to each side. This helps crocodiles and alligators swim more quickly when they are chasing down their dinner. It also allows them to swim underwater at high speed, while barely making a ripple at the surface. This makes it difficult for prey to spot the reptiles swimming toward them—before it is too late!

SNAP AND SNACK!

Adult alligators are rarely attacked, but young alligators often are. Their main predators are turtles and other alligators. A large, stout turtle called an alligator snapping turtle doesn't just snap at alligators–it eats smaller ones, too.

Big Animal, Small Appetite

Crocodiles and alligators do not use energy to keep themselves warm, which means they do not have to eat as much as other animals. For example, an alligator that weighs 800 pounds (362 kg) eats less food than a 100-pound-(45 kg) dog.

Cold-Blooded Killers

Like other reptiles, alligators and crocodiles are cold-blooded, which means that they cannot use their own energy to make heat, but instead take in heat from their surroundings. Alligators are most active and able to hunt when they are warm, so they often bask in the sunlight to warm up their bodies.

If an alligator becomes too hot, it opens its mouth wide! This cools it down, just like panting helps dogs cool down.

Some crocodiles dig caves and burrows to shelter in during weather extremes.

Super Scales

The reptiles' bumpy scales do another very important job. They work like mini solar panels. They collect heat from the sun, which helps the animals gain the energy they need to hunt. Their blood moves from cooler parts of their body into the bumpy scales, where it is heated. Crocodiles and alligators do not need to lie with their entire body in the sun. They can hide mostly underwater and use only the bumps on their head to collect heat. This is very useful when they are lying in wait for prey!

Coping with the Cold

When its surroundings cool down, an alligator slows down, too. When it becomes very cold, the alligator becomes dormant, which means it stops moving around. To make a place in which to rest, the alligator digs out a hole in the bank of a lake or river. It is called a "gator hole." Alligators also dig holes in the mud to keep cool when it is hot. Crocodiles dig holes to keep warm, too. Some larger crocodiles dig very deep holes to sleep in through colder months. Some leave their holes once a day, to bask in the sun before sliding back into the pit. Crocodiles also use this tactic to survive periods of **drought**, when there is little or no rain.

Babies in the Food Chain

Unlike most reptiles, crocodiles and alligators take care of their young. A female alligator lays 20–60 eggs in a nest of rotting plants. A female crocodile lays an average of 10–50 hard-shelled eggs in a nest, depending upon her size and species.

BOYS OR GIRLS?

Nests keep eggs and baby crocodiles and alligators hidden from predators. They also affect whether male or female crocodiles and alligators **hatch** from the eggs. As is the case for many reptiles, the sex of the baby inside the egg is determined by the temperature of the egg in the nest. Warmer temperatures produce more males and cooler temperatures produce more females. If the nest is built on a sunny riverbank, the babies will mostly be males.

Safe Shelters

Alligators and crocodiles make their nests on dry land. They build nests from plants and rotting plant parts, and lay their eggs in them. They then cover them with soil. Some crocodiles dig a hole in the ground and refill it with dirt after they have laid the eggs inside. The mother then stays close to the nest, chasing off any animals that come near.

Baby crocodiles hatch from their eggs.

Crocodile jaws might be fearsome weapons, but they can also carry babies very gently.

Time to Hatch

When baby crocodiles and alligators are ready to hatch, they start to chirp. This is a signal for the mother to uncover the nest, so the babies can then scramble out of their eggs. Once the babies have hatched, the mother carries them very gently in her mouth to the water. The babies are very small when they hatch—only around 9 inches (22 cm) long. Their mother cares for them until they are around a year old.

Young Hunters

Baby alligators and crocodiles can swim and hunt from the moment that they hatch. They stay close to their mother while they hunt, snapping up small fish, water snails, and other little water animals. The babies must be careful, though. Many animals, including turtles, raccoons, snakes, and wading birds, prey on them.

SALTWATER HUNTER

The saltwater crocodile can reach gigantic proportions of 23 feet (7 m) in length and 1 ton (1 mt) in weight. It is the super-sized champion of killer kings.

Saltwater crocodiles, or salties, live in coastal waters such as estuaries or rivers rather than in the ocean. They lie in the water, waiting for any unsuspecting animal to stop for a drink. If they sense prey, they inch closer, moving silently underwater. Then, they burst out, with wide-open jaws that reveal 60 strong teeth, measuring 5 inches (15 cm) long. They seize their prey, clamp on tight, then drag it underwater to drown it. Salties then eat at their leisure.

Saltwater crocodiles have the strongest recorded bite force of any animal on the planet —it is easily powerful enough to crush a human skull.

Salty Sea Trips

Salties live in a large area that stretches from southern India and Sri Lanka, across to the Philippines, and down through Indonesia to northern Australia.

Salties can swim in the ocean, using their long, flat tails to push them forward. These mega-reptiles are more than capable of making long sea trips from one place to another. They may swim hundreds of miles to find a place where they can **breed**. This is because salties are often quite territorial. Males control a patch where they can hunt, and also where they can breed. They fiercely guard these waters, so younger males or males that have lost fights to hold onto their territory have to travel elsewhere to find a territory of their own. Some salties even **migrate** to particular beaches where sea turtle eggs hatch, so they can dine on baby turtles—a favorite snack!

IN THE FOOD CHAIN

Salties are not fussy eaters. Smaller crocs often eat fish and frogs, but the biggest will grab giant prey such as water buffalo. The crocs even kill people, if they venture too close to the crocs!

Saltwater crocodile food chain

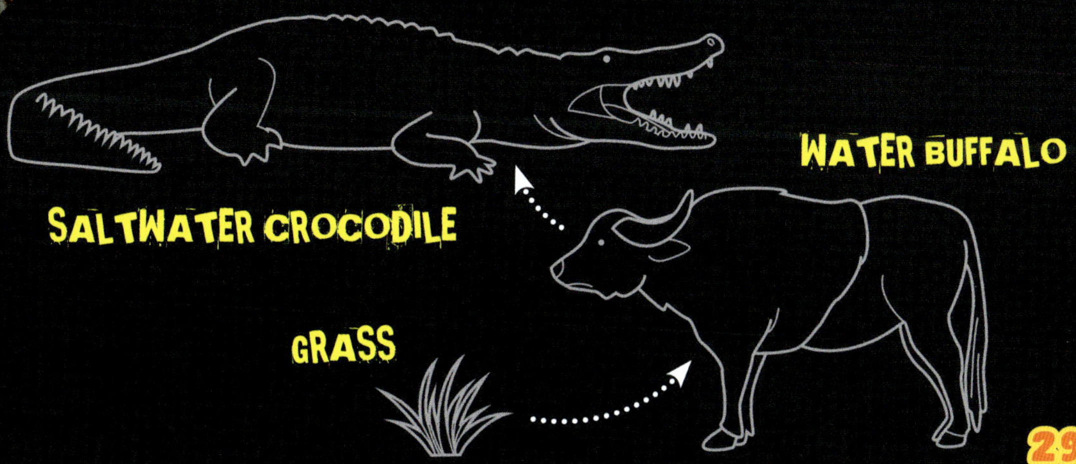

CHAPTER 4

PAST PREDATORS

With their huge, armored bodies and jaws, alligators and crocodiles look like ancient beasts—and they are! The first alligator lived around 35 million years ago, but their ancient relatives have been around since the age of the dinosaurs, around 200 million years ago.

From Land to Water

The first crocodilians were land animals that had huge jaws. Later, they **evolved** and changed into creatures that could swim. Some ran around on land, like dogs, and others lived their whole lives in the open ocean. Some ate plants and others ate meat. These animals then continued to live in and out of water for millions of years. Amazingly, alligators and crocodiles even survived the disaster that wiped out the dinosaurs and other animals 65 million years ago! They probably survived because they could lie low in fresh water and could become dormant in times of cold and drought.

Desmatosuchus *was a gentle giant! The ancient crocodilian fed only on plants.*

SUPER SURVIVOR

Crocodilians have several strengths that may have helped them survive when the dinosaurs could not. Crocodilians are very tough, and can quickly change to suit their environment. They also eat whatever they find, and can even survive for up to a year without eating!

The Link Crocodile

Desmatosuchus means "link crocodile" in the ancient Greek language. It is one of the earliest crocodilians, and lived about 230 million years ago, when it roamed the forests of North America. It measured up to 15 feet (almost 5 m) long and weighed about 1,000 pounds (450 kg). *Desmatosuchus* looked and stood rather like a crocodile and, like a crocodile, it had armored skin.

However, the ancient crocodile relative was much gentler than the crocodiles and alligators of today. It ate only plants and lived in a herd. The herd kept watch for danger while munching greenery. The crocodilian's menacing-looking spikes were not used for attack, but rather for protection from large and hungry predators that lived nearby.

This ancient alligator ***fossil*** *was found in rock. It includes many sharp teeth and long bones that formed the jaws.*

The Carolina Butcher

Carnufex was a terrifying predator that roamed North Carolina about 230 million years ago. This ancient crocodilian walked on its two back legs and at 9 feet (3 m) tall, towered above most of its prey. It waded through ancient mud, **stalking** its prey. It ate small mammals, but its mighty teeth could also pierce the armor of scaly reptiles. It was so deadly it has earned the nickname the "Carolina Butcher."

Blades for Biting

Baurusuchus was an ancient crocodilian that bit into its prey with blade-like teeth. The monster would have **ambushed** its prey, biting deeply and pulling back with its jagged, blade-like teeth to rip its victim's flesh. Then it probably let go, and waited for its prey to die of blood loss before eating it.

Carnufex *was a top predator before dinosaurs roamed North America.*

Tearing Lizard

The name *Dakosaurus* means "tearing lizard," and this **prehistoric** crocodilian lived up to its name. The large, meat-eating predator lived 150–130 million years ago, and spent much, if not all, of its life in the ocean. Its huge, powerful jaws meant it had a very strong bite. That would have allowed it to rip into and feed on any large prey animals that were unfortunate enough to swim near it. *Dakosaurus* grew up to 16 feet (5 m) long, and its torpedo-shaped body helped it swim through the water at speed to chase down its prey.

This is the skull of a *Dakosaurus*.

Mega Hunter

Between 83–72 million years ago, *Deinosuchus* was the largest predator in North America. It grew to about 36 feet (11 m) long and weighed up to 15,000 pounds (7 mt). That's as heavy as a large elephant and as long as a big school bus! Like modern-day alligators, *Deinosuchus* preyed on a variety of animals. It ate sea turtles, dinosaurs, and probably any animal that came to the water's edge to drink. It trapped prey in huge, blunt, and heavy teeth, which were the size of bananas!

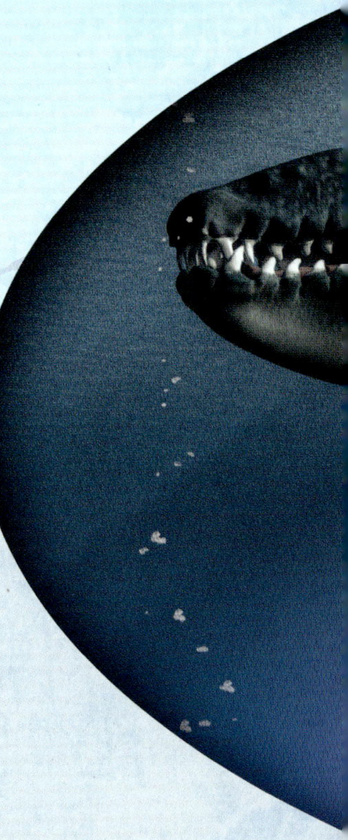

This is a reconstructed Deinosuchus *skeleton. Fossil bones of* Tyrannosaurus rex *dinosaurs have been found with* Deinosuchus *bite marks in them. This proves the ancient alligator attacked the giant dinosaur.*

Mosasaurus *looked a little like a shark and an alligator!*

Hunter of the Seas

Mosasaurus was an unusual crocodilian that lived 70–65 million years ago. It had the head of an alligator, but it also had flipper-like arms and a fin on its tail, like a shark. It swam mainly by moving the large fin on its tail from side to side. Its torpedo-shaped and sleek body helped it speed through the water quickly in hot pursuit of prey such as **ammonites**, bony fish, sea turtles, and even sea birds.

Double Dog Tooth

At about 4 feet (1.5 m) long, *Diplocynodon* was not a large prehistoric crocodilian, but it was still fearsome. Its name means "double dog tooth," and describes the two large, fang-like teeth that extended from its mouth and made it a formidable hunter. *Diplocynodon* lived around 80–70 million years ago in Europe and parts of North America.

DINOSAUR HUNTER

When we think of the age of the dinosaurs, it's the *Tyrannosaurus rex* that looks most terrifying. Yet, even this dangerous giant feared the ancient crocodilian *Sarcosuchus*. This beast was twice as heavy as the largest tyrannosaurs at the time it lived, and its huge jaws were deadly weapons.

Sarcosuchus could grow to 40 feet (12.19 m) long. It lived about 110 million years ago in central Niger, Africa. *Sarcosuchus* was the top predator in the rivers where it lived. It ate anything that came within its range—even large and aggressive dinosaurs. It had 132 huge teeth inside its 6-foot- (1.8 m) long head. It used its powerful jaws to latch onto its prey, and gripped them with its backward-pointing teeth. Then, it dragged its victims into the water to drown them.

Sarcosuchus *had powerful jaws that could easily crush the bones of even large dinosaurs.*

Where in the World?

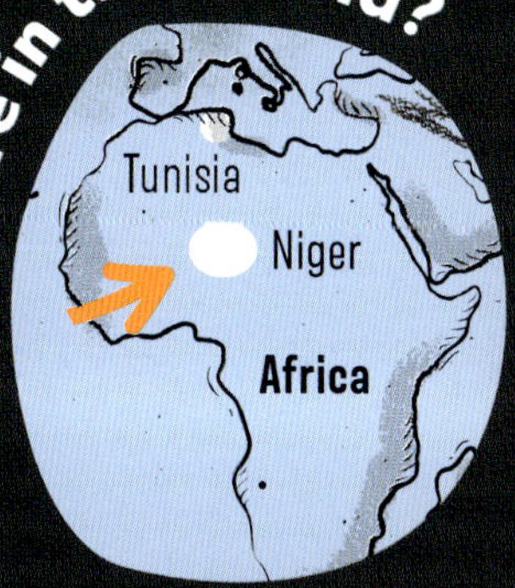

Starved to Death

Most fossils of *Sarcosuchus* have been found in Niger, in Africa. There, *Sarcosuchus* lived in a huge system of rivers. Some fossils of the supercroc have even been found in Tunisia, in North Africa. This means that the giant may have hunted across a huge area.

Sarcosuchus became **extinct** around 115 to 95 million years ago. Most scientists believe it died out due to a lack of food. The supercroc vanished around the same time as the dinosaurs, and like meat-eating dinosaurs, it probably died out because its large prey disappeared.

IN THE FOOD CHAIN

Sarcosuchus fed mainly on large, bony fish that swam in the rivers where it lived. If a larger meal came along, it would have lunged to grab it, feasting on dinosaurs as large as the mighty *Tyrannosaurus rex.*

Sarcosuchus *food chain*

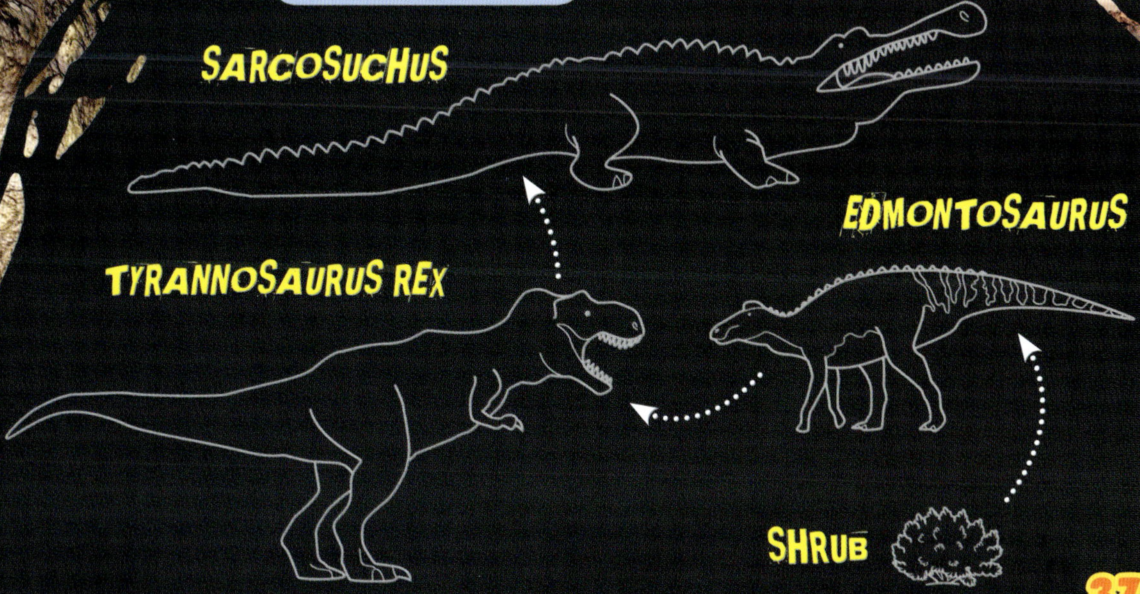

CHAPTER 5

FOOD CHAIN FIGHT

In the 1960s, American alligators became endangered because people hunted and killed them for their skin. In the 1970s and 1980s, laws were passed to protect alligators. Since then, American alligators have recovered so well that there are now more than 1 million alligators in the United States.

A sign in Everglades National Park tells visitors that alligators can be dangerous.

No More Killing

In the past, alligator skins were sold for high prices. They were made into expensive goods, such as purses and shoes. In the United States today, it is illegal to kill an alligator, or to buy or sell its skin. The saltwater and **freshwater** crocodiles of northern Australia are also thriving today thanks to the ban on hunting them, which was introduced in 1970.

One reason Siamese crocodiles are in danger of dying out is that they have been killed to make footwear, bags, belts, wallets, and many other products.

Almost Gone

Chinese alligators are an endangered species. Fewer than 150 of them still live in the wild, living in just a few ponds along the banks of the Yangtze River in China. Local people hunt them because they say the alligators eat too many fish. Most Chinese alligators now live in **captivity**, in zoos around the world and in special centers where they can breed.

Under Threat

More than half of the different species of crocodiles in the world is also under threat. Their habitat is being destroyed, as people drain their land and take it over for their own use. Crocodiles are also being hunted or accidentally caught and drowned in fishing nets. The gharial is an unusual-looking species with a long nose. It lives in India, and is suffering from the destruction of its habitat and being caught in nets.

Survive and Thrive

You might think that removing crocodiles and alligators would help other animals in the wetlands. In fact, the opposite is true, because top predators keep the whole habitat healthy. People have discovered that protecting alligators and crocodiles keeps food chains in balance and helps preserve their habitats.

Alligator Friend

When alligators almost became extinct in wetlands, the numbers of muskrats and nutrias in wetland areas grew. Just one female muskrat alone can have 24 babies in a year. As the numbers of nutrias and muskrats increased, so did the amount of river plants they ate. River plants help wetlands, because they are the basis of wetland food chains. They filter the water and keep it clean, and stop the water washing away.

Signs warn people where it is not safe to swim because of crocodiles.

NOT A TRUE THREAT

People threaten alligators, but alligators and crocodiles rarely attack human beings, if people keep their distance. In Florida, the US state that has the most alligators, only five people were killed between 1973 and 1990. In South Carolina, only 11 people have reported being bitten by an alligator since 1948, and no one has been killed.

Philippine crocodiles eat unwanted crop pests such as snails and rats.

By eating too many plants, nutrias and muskrats damaged the wetland. After alligators were reintroduced to wetlands, they preyed on the muskrats and nutrias and kept their numbers in check.

Farmer's Helper

The Philippine crocodile helps farmers grow food for people! This 10-foot-(3 m) long crocodile is shy and avoids humans, but loves to eat small animals such as golden apple snails and rats. These two animals are considered pests by farmers, because they eat their rice crops. By eating the pests, the crocodile helps the farmers.

Water Store

Gator holes made by alligators and the burrows and nests built by crocodiles also help wetlands. They are places where plants can grow and animals can shelter when it is too hot or too cold. Gator and crocodile holes also hold and store water. When a drought or dry season happens, and there is less rainwater available, a gator hole water source can save the lives of other animals.

FISH HUNTER

With its distinctive long, narrow snout, and widely set eyes, there is no mistaking the gharial. Although it can reach 20 feet (6 m) in length, the only threat this gentle giant poses is to fish!

Gharials are specialist fisher kings. Their jaws contain up to 110 razor-sharp, hook-shaped pointed teeth. They use the teeth to bite into fast-moving, slippery prey. The teeth interlock when the jaws are closed, so that small fish cannot wriggle away through any gaps. Gharials have excellent eyesight, which they use to hunt fish underwater. Their eyes act a little like night-vision goggles in murky or dark water, helping the gharials to see clearly. They use their flat tail and webbed feet to swim quickly through water.

A gharial's snout is long, and could easily drag through the water. To stop that, the gharial moves the snout up, down, or sideways. That stops the snout slowing down the gharial as it speeds toward its prey.

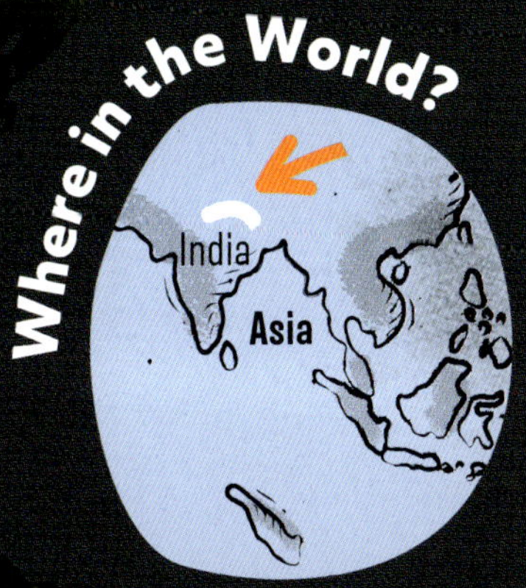

River Problems

Gharials live in rivers and reservoirs in eastern Asia, in north India and Nepal. Mostly, these habitats are in protected areas such as national parks, because gharials are endangered.

However, in unprotected areas, people dam rivers to **irrigate** farmland and produce **hydropower**, which disturbs riverbanks where gharials nest. That stops them from breeding. Blocking rivers also changes the flow of water. Without connections between one river and the next, gharials cannot move around to find mates to breed with. When people pollute rivers or overfish, it leaves fewer fish for gharials and other animals, such as river dolphins, to eat. Gharials can also become caught in the fishing nets, and drown.

IN THE FOOD CHAIN

Young gharials snack on insects, but as they grow, they eat only fish, some as big as large catfish.

Gharial food chain

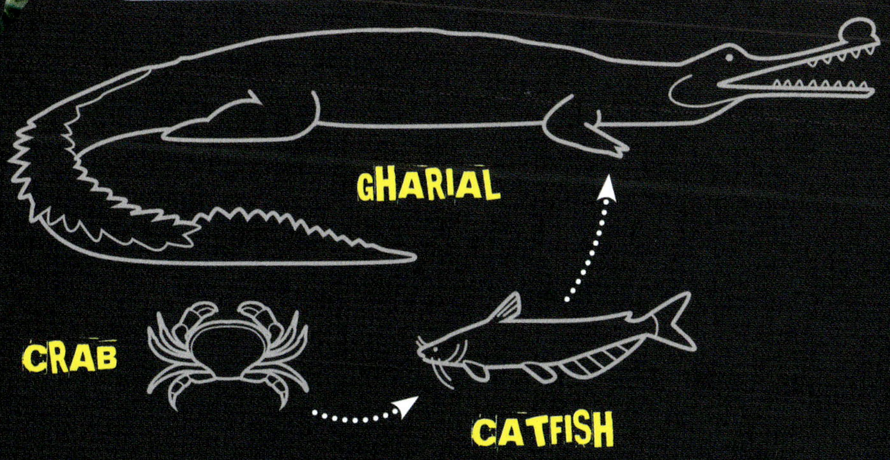

ULTIMATE KILLER!

WHO IS THE ULTIMATE KILLER KING OF THE CROCODILE AND ALLIGATOR KINGDOM TODAY? CHECK OUT THE FOOD CHAIN FACTS BELOW, THEN YOU DECIDE!

NILE CROCODILE

AVERAGE LENGTH	16 feet (4.8 m)
FASTEST SPEED (IN WATER)	19–22 miles per hour (30–35 km/h)
FOOD CHAIN SNACK	fish
FOOD CHAIN FEAST	wildebeest
KILLER BLOW	has the deadliest bite of any alligator or crocodile

AMERICAN ALLIGATOR

AVERAGE LENGTH	11.2 feet (3.4 m)
FASTEST SPEED (IN WATER)	20 miles per hour (32 km/h)
FOOD CHAIN SNACK	fish
FOOD CHAIN FEAST	raccoon
KILLER BLOW	will eat just about anything!

SALTWATER CROCODILE

AVERAGE LENGTH	17 feet (5 m)
FASTEST SPEED (IN WATER)	15-18 miles per hour (24-29 km/h)
FOOD CHAIN SNACK	frog
FOOD CHAIN FEAST	water buffalo
KILLER BLOW	has 60 strong teeth, each measuring 5 inches (15 cm) long

GHARIAL

AVERAGE LENGTH	11-23 feet (3.5-7 m)
FASTEST SPEED (IN WATER)	15 miles per hour (24 km/h)
FOOD CHAIN SNACK	fish
FOOD CHAIN FEAST	duck
KILLER BLOW	has more than 100 needle-sharp teeth

CROWN ME!

GLOSSARY

ambushed made a surprise attack from a hidden place

ammonites ancient, shelled, water-living creatures

breed to mate and then produce babies

camouflaged hidden against a background

captivity being confined or caged

climate change the change in temperature and weather over a long period of time

drought a period with little or no rainfall

efficiently working well and without wasting energy

endangered at risk of dying out

environment the natural place in which a plant or animal lives

estuarine describes a sandy or muddy area where a river meets an ocean or sea

evolved changed over time to better suit its environment

extinct died out

fossil the hardened remains of a dead plant or animal

freshwater water that is found in streams, rivers, and lakes. Fresh water does not contain salt

habitats places in which animals and plants make their homes

hatch to break out of an egg

hydropower a system of making electricity that uses moving water

irrigate to supply water for growing plants

mammals animals that give birth to babies and feed them with milk from their bodies

migrate to move from one region to another, often to find food or breed

pollution the introduction of harmful materials into the environment

predators animals that hunt and eat other animals

prehistoric describes a time before people began to record things, or write them down

prey animals that are hunted and eaten by other animals

range the area where an animal spends its time

snout the mouth and nose area of an animal

species a group of living creatures that have similar characteristics and can breed with each other

stalking quietly following a prey animal

swamps watery areas with many plants

territory area of land or sea that an animal defends as its own

tropical describes an area that is near to the tropics, which are areas toward the middle of Earth

vibrations movements that can be felt

webbed having skin between the fingers or toes

wetland land that is covered with water and plants

FIND OUT MORE

Books

London, Martha. *Saving American Crocodiles* (Saving Animals). North Star Editions, 2021.

Orr, Tamra. *Crocodile or Alligator?* (21st Century Junior Library: Which Is Which?). Cherry Lake Publishing, 2019.

Shield, Jennifer. *Alligator or Crocodile?: A Compare and Contrast Book*. Arbordale Publishing, 2023.

Websites

Learn more fascinating facts about crocodiles at:
https://easyscienceforkids.com/crocodile-facts-for-kids

Find out more about the alligator killer kings of the wetlands at:
https://kids.britannica.com/kids/article/alligator/352737

Discover more about the differences between crocodiles and alligators at:
www.sciencekids.co.nz/sciencefacts/animals/crocodilealligatordifferences.html

Publisher's note to educators and parents:
All the websites featured above have been carefully reviewed to ensure that they are suitable for students. However, many websites change often, and we cannot guarantee that a site's future contents will continue to meet our high standards of educational value. Please be advised that students should be closely monitored whenever they access the Internet.

INDEX

About the Author

Katherine Eason has written many books about animals and their environments. In writing and researching this book, she has discovered that crocodiles and alligators are lethal predators that deserve our respect and protection.